On the

Translated by Isabel F. Hapgood

CHAPTER I

. . . {1} The justification of all persons who have freed themselves from toil is now founded on experimental, positive science. The scientific theory is as follows:-

"For the study of the laws of life of human societies, there exists but one indubitable method,--the positive, experimental, critical method

"Only sociology, founded on biology, founded on all the positive sciences, can give us the laws of humanity. Humanity, or human communities, are the organisms already prepared, or still in process of formation, and which are subservient to all the laws of the evolution of organisms.

"One of the chief of these laws is the variation of destination among the portions of the organs. Some people command, others obey. If some have in superabundance, and others in want, this arises not from the will of God, not because the empire is a form of manifestation of personality, but because in societies, as in organisms, division of labor becomes indispensable for life as a whole. Some people perform the muscular labor in societies; others, the mental labor."

Upon this doctrine is founded the prevailing justification of our time.

Not long ago, their reigned in the learned, cultivated world, a moral philosophy, according to which it appeared that every thing which exists is reasonable; that there is no such thing as evil or good; and that it is unnecessary for man to war against evil, but that it is only necessary for him to display intelligence,--one man in the military service, another in the judicial, another on the violin. There have been many and varied expressions of human wisdom, and these phenomena were known to the men of the nineteenth century. The wisdom of Rousseau and of Lessing, and Spinoza and Bruno, and all the wisdom of antiquity; but no one man's wisdom overrode the crowd. It was impossible to say even this,--that Hegel's success was the result of the symmetry of this theory. There were other equally symmetrical theories,--those of Descartes, Leibnitz, Fichte,

Schopenhauer. There was but one reason why this doctrine won for itself, for a season, the belief of the whole world; and this reason was, that the deductions of that philosophy winked at people's weaknesses. These deductions were summed up in this,--that every thing was reasonable, every thing good; and that no one was to blame.

When I began my career, Hegelianism was the foundation of every thing. It was floating in the air; it was expressed in newspaper and periodical articles, in historical and judicial lectures, in novels, in treatises, in art, in sermons, in conversation. The man who was not acquainted with Hegal had no right to speak. Any one who desired to understand the truth studied Hegel. Every thing rested on him. And all at once the forties passed, and there was nothing left of him. There was not even a hint of him, any more than if he had never existed. And the most amazing thing of all was, that Hegelianism did not fall because some one overthrew it or destroyed it. No! It was the same then as now, but all at once it appeared that it was of no use whatever to the learned and cultivated world.

There was a time when the Hegelian wise men triumphantly instructed the masses; and the crowd, understanding nothing, blindly believed in every thing, finding confirmation in the fact that it was on hand; and they believed that what seemed to them muddy and contradictory there on the heights of philosophy was all as clear as the day. But that time has gone by. That theory is worn out: a new theory has presented itself in its stead. The old one has become useless; and the crowd has looked into the secret sanctuaries of the high priests, and has seen that there is nothing there, and that there has been nothing there, save very obscure and senseless words. This has taken place within my memory.

"But this arises," people of the present science will say, "from the fact that all that was the raving of the theological and metaphysical period; but now there exists positive, critical science, which does not deceive, since it is all founded on induction and experiment. Now our erections are not shaky, as they formerly were, and only in our path lies the solution of all the problems of humanity."

But the old teachers said precisely the same, and they were no fools; and we know that there were people of great intelligence among them. And precisely thus, within my memory, and with no less confidence, with no less recognition on the part of the crowd of so- called cultivated people, spoke the Hegelians. And neither were our Herzens, our Stankevitches, or our Byelinskys fools. But whence arose that marvellous manifestation, that sensible people should preach with the greatest assurance, and that the crowd should accept with devotion, such unfounded and unsupportable teachings? There is but one reason,--that the teachings thus inculcated justified people in their evil life.

A very poor English writer, whose works are all forgotten, and recognized as the most insignificant of the insignificant, writes a treatise on population, in which he devises a fictitious law concerning the increase of population disproportionate to the means of subsistence. This fictitious law, this writer encompasses with mathematical formulae founded on nothing whatever; and then he launches it on the world. From the frivolity and the stupidity of this hypothesis, one would suppose that it would not attract the attention of any one, and that it would sink into oblivion, like all the works of the same author which followed it; but it turned out quite otherwise. The hack-writer who penned this treatise instantly becomes a scientific authority, and maintains himself upon that height for nearly half a century. Malthus! The Malthusian theory,- -the law of the increase of the population in geometrical, and of the means of subsistence in arithmetical proportion, and the wise and natural means of restricting the population,--all these have become scientific, indubitable truths, which have not been confirmed, but which have been employed as axioms, for the erection of false theories. In this manner have learned and cultivated people proceeded; and among the herd of idle persons, there sprung up a pious trust in the great laws expounded by Malthus. How did this come to pass? It would seem as though they were scientific deductions, which had nothing in common with the instincts of the masses. But this can only appear so for the man who believes that science, like the Church, is something self-contained, liable to no errors, and not simply the imaginings of weak and erring folk, who merely substitute the imposing word "science," in place of the thoughts and words of the people, for the sake of impressiveness.

All that was necessary was to make practical deductions from the theory of Malthus, in order to perceive that this theory was of the most human sort, with the best defined of objects. The deductions directly arising from this theory were the following: The wretched condition of the laboring classes was such in accordance with an unalterable law, which does not depend upon men; and, if any one is to blame in this matter, it is the hungry laboring classes themselves. Why are they such fools as to give birth to children, when they know that there will be nothing for the children to eat? And so this deduction, which is valuable for the herd of idle people, has had this result: that all learned men overlooked the incorrectness, the utter arbitrariness of these deductions, and their insusceptibility to proof; and the throng of cultivated, i.e., of idle people, knowing instinctively to what these deductions lead, saluted this theory with enthusiasm, conferred upon it the stamp of truth, i.e., of science, and dragged it about with them for half a century.

Is not this same thing the cause of the confidence of men in positive critical-experimental science, and of the devout attitude of the crowd towards that which it preaches? At first it seems strange, that the theory of evolution can in any manner justify people in their evil ways; and it seems as though the scientific theory of evolution has to deal only with facts, and that it does nothing else but observe facts.

But this only appears to be the case.

Exactly the same thing appeared to be the case with the Hegelian doctrine, in a greater degree, and also in the special instance of the Malthusian doctrine. Hegelianism was, apparently, occupied only with its logical constructions, and bore no relation to the life of mankind. Precisely this seemed to be the case with the Malthusian theory. It appeared to be busy itself only with statistical data. But this was only in appearance.

Contemporary science is also occupied with facts alone: it investigates facts. But what facts? Why precisely these facts, and no others?

The men of contemporary science are very fond of saying, triumphantly and confidently, "We investigate only facts," imagining that these words contain some meaning. It is impossible to investigate facts alone, because the facts which are subject to our investigation are INNUMERABLE (in the definite sense of that word),- -innumerable. Before we proceed to investigate facts, we must have a theory on the foundation of which these or those facts can be inquired into, i.e., selected from the incalculable quantity.

And this theory exists, and is even very definitely expressed, although many of the workers in contemporary science do not know it, or often pretend that they do not know it. Exactly thus has it always been with all prevailing and guiding doctrines. The foundations of every doctrine are always stated in a theory, and the so-called learned men merely invent further deductions from the foundations once stated. Thus contemporary science is selecting its facts on the foundation of a very definite theory, which it sometimes knows, sometimes refuses to know, and sometimes really does not know; but the theory exists.

The theory is as follows: All mankind is an undying organism; men are the particles of that organism, and each one of them has his own special task for the service of others. In the same manner, the cells united in an organism share among them the labor of fight for existence of the whole organism; they magnify the power of one capacity, and weaken another, and unite in one organ, in order the better to supply the requirements of the whole organism. And exactly in the same manner as with gregarious animals,--ants or bees,--the separate individuals divide the labor among them. The queen lays the egg, the drone fructifies it; the bee works his whole life long. And precisely this thing takes place in mankind and in human societies. And therefore, in order to find the law of life for man, it is necessary to study the laws of the life and the development of organisms.

In the life and development of organisms, we find the following laws: the law of differentiation and integration, the law that every phenomenon is accompanied not by direct consequences alone, another law regarding the instability of type, and so on. All this seems very innocent; but it is only necessary to draw the deductions from all these laws, in order to immediately perceive that these laws incline in the same direction as the law of Malthus. These laws all point to one thing; namely, to the recognition of that division of labor which exists in human communities, as organic, that is to say, as indispensable. And therefore, the unjust position in which we, the people who have freed ourselves from labor, find ourselves, must be regarded not from the point of view of common-sense and justice, but merely as an undoubted fact, confirming the universal law.

Moral philosophy also justified every sort of cruelty and harshness; but this resulted in a philosophical manner, and therefore wrongly. But with science, all this results scientifically, and therefore in a manner not to be doubted.

How can we fail to accept so very beautiful a theory? It is merely necessary to look upon human society as an object of contemplation; and I can console myself with the thought that my activity, whatever may be its nature, is a functional activity of the organism of humanity, and that therefore there cannot arise any question as to whether it is just that I, in employing the labor of others, am doing only that which is agreeable to me, as there can arise no question as to the division of labor between the brain cells and the muscular cells. How is it possible not to admit so very beautiful a theory, in order that one may be able, ever after, to pocket one's conscience, and have a perfectly unbridled animal existence, feeling beneath one's self that support of science which is not to be shaken nowadays!

And it is on this new doctrine that the justification for men's idleness and cruelty is now founded.

CHAPTER II

.This doctrine had its rise not so very long--fifty years--ago. Its principal founder was the French savant Comte. There occurred to Comte,--a systematist, and a religious man to boot,--under the influence of the then novel physiological investigations of Biche, the old idea already set forth by Menenius Agrippa,--the idea that human society, all humanity even, might be regarded as one whole, as an organism; and men as living parts of the separate organs, having each his own definite appointment to serve the entire organism.

This idea so pleased Comte, that upon it he began to erect a philosophical theory; and this theory so carried him away, that he utterly forgot that the point of departure for his theory was nothing more than a very pretty comparison, which was suitable for a fable, but which could by no means serve as the foundation for science. He, as frequently happens, mistook his pet hypothesis for an axiom, and imagined that his whole theory was erected on the very firmest of foundations. According to his theory, it seemed that since humanity is an organism, the knowledge of what man is, and of what should be his relations to the world, was possible only through a knowledge of the features of this organism. For the knowledge of these qualities, man is enabled to take observations on other and lower organisms, and to draw conclusions from their life. Therefore, in the fist place, the true and only method, according to Comte, is the inductive, and all science is only such when it has experiment as its basis; in the second place, the goal and crown of sciences is formed by that new science dealing with the imaginary organism of humanity, or the super-organic being,--humanity,-- and this newly devised science is sociology.

And from this view of science it appears, that all previous knowledge was deceitful, and that the whole story of humanity, in the sense of self-knowledge, has been divided into three, actually into two, periods: the theological and metaphysical period, extending from the beginning of the world to Comte, and the present period,--that of the only true science, positive science,--beginning with Comte.

All this was very well. There was but one error, and that was this,--that the whole edifice was erected on the sand, on the arbitrary and false assertion that humanity is an organism. This assertion was arbitrary, because we have just as much right to admit the existence of a human organism, not subject to observation, as we have to admit the existence of any other invisible, fantastic being. This assertion was erroneous, because for the understanding of humanity, i.e., of men, the definition of an organism was incorrectly constructed, while in humanity itself all actual signs of organism,--the centre of feeling or consciousness, are lacking. {2}

But, in spite of the arbitrariness and incorrectness of the fundamental assumption of positive philosophy, it was accepted by the so-called cultivated world with the greatest sympathy. In this connection, one thing is worthy of note: that out of the works of Comte, consisting of two parts, of positive philosophy and of positive politics, only the first was adopted by the learned world,- -that part which justifieth, on new promises, the existent evil of human societies; but the second part, treating of the moral obligations of altruism, arising from the recognition of mankind as an organism, was regarded as not only of no importance, but as trivial and unscientific. It was a repetition of the same thing that had happened in the case of Kant's works. The "Critique of Pure Reason" was adopted by the scientific crowd; but the "Critique of Applied Reason," that part which contains the gist of moral doctrine, was repudiated. In Kant's doctrine, that was accepted as scientific which subserved the existent evil. But the positive philosophy, which was accepted by the crowd, was founded on an arbitrary and erroneous basis, was in itself too unfounded, and therefore unsteady, and could not support itself alone. And so, amid all the multitude of the idle plays of thought of the men professing the so-called science, there presents itself an assertion equally devoid of novelty, and equally arbitrary and erroneous, to the effect that living beings, i.e., organisms, have had their rise in each other,--not only one organism from another, but one from many; i.e., that in a very long interval of time (in a million of years, for instance), not only could a duck and a fish proceed from one ancestor, but that one animal might result from a whole hive of bees. And this arbitrary and erroneous assumption was accepted by the learned world with still greater and more universal sympathy. This assumption was arbitrary, because no one has ever seen how one organism is made from another, and therefore the hypothesis as to the origin of species will always remain an hypothesis, and not an experimental fact. And this hypothesis was also erroneous, because the decision of the question as to the origin of species-- that they have originated, in consequence of the law of heredity and fitness, in the course of an interminably long time--is no solution at all, but merely a re-statement of the problem in a new form.

According to Moses' solution of the question (in the dispute with whom the entire significance of this theory lies), it appears that the diversity of the species of living creatures proceeded according to the will of God, and according to His almighty power; but according to the theory of evolution, it appears that the difference between living creatures arose by chance, and on account of varying conditions of heredity and surroundings, through an endless period of time. The theory of evolution, to speak in simple language, merely asserts, that by chance, in an incalculably long period of time, out of any thing you like, any thing else that you like may develop.

This is no answer to the problem. And the same problem is differently expressed: instead of will, chance is offered, and the co-efficient of the eternal is transposed from the power to the time. But this fresh assertion strengthened Comte's assertion. And, moreover, according to the ingenuous confession of the founder of Darwin's theory himself, his idea was aroused in him by the law of Malthus; and he therefore propounded the theory of the struggle of living creatures and people for existence, as the fundamental law of every living thing. And lo! only this was needed by the throng of idle people for their justification.

Two insecure theories, incapable of sustaining themselves on their feet, upheld each other, and acquired the semblance of stability. Both theories bore with them that idea which is precious to the crowd, that in the existent evil of human societies, men are not to blame, and that the existing order of things is that which should prevail; and the new theory was adopted by the throng with entire faith and unheard-of enthusiasm. And behold, on the strength of these two arbitrary and erroneous hypotheses, accepted as dogmas of belief, the new scientific doctrine was ratified.

Spencer, for example, in one of his first works, expresses this doctrine thus:-

"Societies and organisms," he says, "are alike in the following points:-

"1. In that, beginning as tiny aggregates, they imperceptibly grow in mass, so that some of them attain to the size of ten thousand times their original bulk.

"2. In that while they were, in the beginning, of such simple structure, that they can be regarded as destitute of all structure, they acquire during the period of their growth a constantly increasing complication of structure.

"3. In that although in their early, undeveloped period, there exists between them hardly any interdependence of parts, their parts gradually acquire an interdependence, which eventually becomes so strong, that the life and activity of each part becomes possible only on condition of the life and activity of the remaining parts.

"4. In that life and the development of society are independent, and more protracted than the life and development of any one of the units constituting it, which are born, grow, act, reproduce themselves, and die separately; while the political body formed from them, continues to live generation after generation, developing in mass in perfection and functional activity."

The points of difference between organisms and society go farther; and it is proved that these differences are merely apparent, but that organisms and societies are absolutely similar.

For the uninitiated man the question immediately presents itself: "What are you talking about? Why is mankind an organism, or similar to an organism?"

You say that societies resemble organisms in these four features; but it is nothing of the sort. You only take a few features of the organism, and beneath them you range human communities. You bring forward four features of resemblance, then you take four features of dissimilarity, which are, however, only apparent (according to you); and you thence conclude that human societies can be regarded as organisms. But surely, this is an empty game of dialectics, and nothing more. On the same foundation, under the features of an organism, you may range whatever you please. I will take the fist thing that comes into my head. Let us suppose it to be a forest,-- the manner in which it sows itself in the plain, and spreads abroad. 1. Beginning with a small aggregate, it increases imperceptibly in mass, and so forth. Exactly the same thing takes place in the fields, when they gradually seed themselves down, and bring forth a forest. 2. In the beginning the structure is simple: afterwards it increases in complication, and so forth. Exactly the same thing happens with the forest,--in the first place, there were only bitch- trees, then came brush-wood and hazel-bushes; at first all grow erect, then they interlace their branches. 3. The interdependence of the parts is so augmented, that the life of each part depends on the life and activity of the remaining parts. It is precisely so with the forest,--the hazel-bush warms the tree-boles (cut it down, and the other trees will freeze), the hazel-bush protects from the wind, the seed-bearing trees carry on reproduction, the tall and leafy trees afford shade, and the life of one tree depends on the life of another. 4. The separate parts may die, but the whole lives. Exactly the case with the forest. The forest does not mourn one tree.

Having proved that, in accordance with this theory, you may regard the forest as an organism, you fancy that you have proved to the disciples of the organic doctrine the error of their definition. Nothing of the sort. The definition which they give to the organism is so inaccurate and so elastic that under this definition they may include what they will. "Yes," they say; "and the forest may also be regarded as an organism. The forest is mutual re-action of individuals, which do not annihilate each other,--an aggregate; its parts may also enter into a more intimate union, as the hive of bees constitutes itself an organism." Then you will say, "If that is so, then the birds and the insects and the grass of this forest, which re-act upon each other, and do not destroy each other, may also be regarded as one organism, in company with the trees." And to this also they will agree. Every collection of living individuals, which re-act upon each other, and do not destroy each other, may be regarded as organisms, according to their theory. You may affirm a connection and interaction between whatever you choose, and, according to evolution, you may affirm, that, out of whatever you please, any other thing that you please may proceed, in a very long period of time.

And the most remarkable thing of all is, that this same identical positive science recognizes the scientific method as the sign of true knowledge, and has itself defined what it designates as the scientific method.

By the scientific method it means common-sense.

And common-sense convicts it at every step. As soon as the Popes felt that nothing holy remained in them, they called themselves most holy.

As soon as science felt that no common-sense was left in her she called herself sensible, that is to say, scientific science.

CHAPTER III

.Division of labor is the law of all existing things, and, therefore, it should be present in human societies. It is very possible that this is so; but still the question remains, Of what nature is that division of labor which I behold in my human society? is it that division of labor which should exist? And if people regard a certain division of labor as unreasonable and unjust, then no science whatever can convince men that that should exist which they regard as unreasonable and unjust.

Division of labor is the condition of existence of organisms, and of human societies; but what, in these human societies, is to be regarded as an organic division of labor? And, to whatever extent science may have investigated the division of labor in the cells of worms, all these observations do not compel a man to acknowledge that division of labor to be correct which his own sense and conscience do not recognize as correct. No matter how convincing may be the proofs of the division of labor of the cells in the organisms studied, man, if he has not parted with his judgment, will say, nevertheless, that a man should not weave calico all his life, and that this is not division of labor, but persecution of the people. Spencer and others say that there is a whole community of weavers, and that the profession of weaving is an organic division of labor. There are weavers; so, of course, there is such a division of labor. It would be well enough to speak thus if the colony of weavers had arisen by the free will of its member's; but we know that it is not thus formed of their initiative, but that we make it. Hence it is necessary to find out whether we have made these weavers in accordance with an organic law, or with some other.

Men live. They support themselves by agriculture, as is natural to all men. One man has set up a blacksmith's forge, and repaired his plough; his neighbor comes to him, and asks him to mend his also, and promises him in return either work or money. A third comes, and a fourth; and in the community formed by these men, there arises the following division of labor,--a blacksmith is created. Another man has instructed his children well; his neighbor brings his children to him, and requests him to teach them also, and a teacher is created. But both blacksmith and teacher have been created, and continue to be such, merely because they have been asked; and they remain such as long as they are requested to be blacksmith and teacher. If it should come to pass that many blacksmiths and teachers should set themselves up, or that their work is not requited, they will immediately, as commonsense demands and as always happens when there is no occasion for disturbing the regular course of division of labor,--they will immediately abandon their trade, and betake themselves once more to agriculture.

Men who behave thus are guided by their sense, their conscience; and hence we, the men endowed with sense and conscience, all assert that such a division of labor is right. But if it should chance that the blacksmiths were able to compel other people to work for them, and should continue to make horse-shoes when they were not wanted, and if the teachers should go on teaching when there was no one to teach, then it is obvious to every sane man, as a man, i.e., as a being endowed with reason and conscience, that this would not be division, but appropriation, of labor. And yet precisely that sort of activity is what is called division of labor by scientific science. People do that which others do not think of requiring, and demand that they shall be supported for so doing, and say that this is just because it is division of labor.

That which constitutes the cause of the economical poverty of our age is what the English call over-production (which means that a mass of things are made which are of no use to anybody, and with which nothing can be done).

It would be odd to see a shoemaker, who should consider that people were bound to feed him because he incessantly made boots which had been of no use to any one for a long time; but what shall we say of those men who make nothing,--who not only produce nothing that is visible, but nothing that is of use for people at large,--for whose wares there are no customers, and who yet demand, with the same boldness, on the ground of division of labor, that they shall be supplied with fine food and drink, and that they shall be dressed well? There may be, and there are, sorcerers for whose services a demand makes itself felt, and for this purpose there are brought to them pancakes and flasks; but it is difficult to imagine the existence of sorcerers whose spells are useless to every one, and who boldly demand that they shall be luxuriously supported because they exercise sorcery. And it is the same in our world. And all this comes about on the basis of that false conception of the division of labor, which is defined not by reason and conscience, but by observation, which men of science avow with such unanimity.

Division of labor has, in reality, always existed, and still exists; but it is right only when man decides with his reason and his conscience that it should be so, and not when he merely investigates it. And reason and conscience decide the question for all men very simply, unanimously, and in a manner not to be doubted. They always decide it thus: that division of labor is right only when a special branch of man's activity is so needful to men, that they, entreating him to serve them, voluntarily propose to support him in requital for that which he shall do for them. But, when a man can live from infancy to the age of thirty years on the necks of others, promising to do, when he shall have been taught, something extremely useful, for which no one asks him; and when, from the age of thirty until his death, he can live in the same manner, still merely on the promise to do something, for which there has been no request, this will not be division of labor (and, as a matter of fact, there is no such thing in our society), but it will be what it already is,-- merely the appropriation, by force, of the toil of others; that same appropriation by force of the toil of others which the philosophers formerly designated by various names,--for instance, as indispensable forms of life,--but which scientific science now calls the organic division of labor.

The whole significance of scientific science lies in this alone. It has now become a distributer of diplomas for idleness; for it alone, in its sanctuaries, selects and determines what is parasitical, and what is organic activity, in the social organism. Just as though every man could not find this out for himself much more accurately and more speedily, by taking counsel of his reason and his conscience. It seems to men of scientific science, that there can be no doubt of this, and that their activity is also indubitably organic; they, the scientific and artistic workers, are the brain cells, and the most precious cells in the whole organism.

Ever since men--reasoning beings--have existed, they have distinguished good from evil, and have profited by the fact that men have made this distinction before them; they have warred against evil, and have sought the good, and have slowly but uninterruptedly advanced in that path. And divers delusions have always stood before men, hemming in this path, and having for their object to demonstrate to them, that it was not necessary to do this, and that it was not necessary to live as they were living. With fearful conflict and difficulty, men have freed themselves from many delusions. And behold, a new and a still more evil delusion has sprung up in the path of mankind,--the scientific delusion.

This new delusion is precisely the same in nature as the old ones; its gist lies in secretly leading astray the activity of our reason and conscience, and of those who have lived before us, by something external. In scientific science, this external thing is-- investigation.

The cunning of this science consists in this,--that, after pointing out to men the coarsest false interpretations of the activity of the reason and conscience of man, it destroys in them faith in their own reason and conscience, and assures them that every thing which their reason and conscience say to them, that all that these have said to the loftiest representatives of man heretofore, ever since the world has existed,--that all this is conventional and subjective. "All this must be abandoned," they say; "it is impossible to understand the truth by the reason, for we may be mistaken. But there exists another unerring and almost mechanical path: it is necessary to investigate facts."

But facts must be investigated on the foundation of scientific science, i.e., of the two hypotheses of positivism and evolution, which are not borne out by any thing, and which give themselves out as undoubted truths. And the reigning science announces, with delusive solemnity, that the solution of all problems of life is possible only through the study of facts, of nature, and, in particular, of organisms. The credulous mass of young people, overwhelmed by the novelty of this authority, which has not yet been overthrown or even touched by criticism, flings itself into the study of natural sciences, into that sole path, which, according to the assertion of the reigning science, can lead to the elucidation of the problems of life.

But the farther the disciples proceed in this study, the farther and farther does not only the possibility, but even the very idea, of the solution of the problems of life withdraw from them, and the more and more do they become accustomed, not so much to investigate, as to believe in the assertions of other investigators (to believe in cells, in protoplasm, in the fourth condition of bodies, and so forth); the more and more does the form veil the contents from them; the more and more do they lose the consciousness of good and evil, and the capacity of understanding those expressions and definitions of good and evil which have been elaborated through the whole foregoing life of mankind; and the more and more do they appropriate to themselves the special scientific jargon of conventional expressions, which possesses no universally human significance; and the deeper and deeper do they plunge into the debris of utterly unilluminated investigations; the more and more do they lose the power, not only of independent thought, but even of understanding the fresh human thought of others, which lies beyond the bounds of their Talmud. But the principal thing is, that they pass their best years in getting disused to life; they grow accustomed to consider their position as justifiable; and they convert themselves physically into utterly useless parasites, and mentally they dislocate their brains and become mental eunuchs. And in precisely the same manner, according to the measure of their folly, do they acquire self-conceit, which deprives them forever of all possibility of return to a simple life of toil, to a simple, clear, and universally human train of reasoning.

Division of labor always has existed in human communities, and will probably always exist; but the question for us lies not in the fact that it has existed, and that it will exist, but in this,--how are we to govern ourselves so that this division shall be right? But if we take investigation as our rule of action, we by this very act repudiate all rule; then in that case we shall regard as right every division of labor which we shall descry among men, and which appears to us to be right--to which conclusion the prevailing scientific science also leads.

Division of labor!

Some are busied in mental or moral, others in muscular or physical, labor. With what confidence people enunciate this! They wish to think so, and it seems to them that, in point of fact, a perfectly regular exchange of services does take place.

But we, in our blindness, have so completely lost sight of the responsibility which we have assumed, that we have even forgotten in whose name our labor is prosecuted; and the very people whom we have undertaken to serve have become the objects of our scientific and artistic activity. We study and depict them for our amusement and diversion. We have totally forgotten that what we need to do is not to study and depict them, but to serve them. To such a degree have we lost sight of this duty which we have taken upon us, that we have not even noticed that what we have undertaken to perform in the realm of science and art has been accomplished not by us, but by others, and that our place has turned out to be occupied.

It proves that while we have been disputing, one about the spontaneous origin of organisms, another as to what else there is in protoplasm, and so on, the common people have been in need of spiritual food; and the unsuccessful and rejected of art and science, in obedience to the mandate of adventurers who have in view the sole aim of profit, have begun to furnish the people with this spiritual food, and still so furnish them. For the last forty years in Europe, and for the last ten years with us here in Russia, millions of books and pictures and song-books have been distributed, and stalls have been opened, and the people gaze and sing and receive spiritual nourishment, but not from us who have undertaken to provide it; while we, justifying our idleness by that spiritual food which we are supposed to furnish, sit by and wink at it.

But it is impossible for us to wink at it, for our last justification is slipping from beneath our feet. We have become specialized. We have our particular functional activity. We are the brains of the people. They support us, and we have undertaken to teach them. It is only under this pretence that we have excused ourselves from work. But what have we taught them, and what are we now teaching them? They have waited for years--for tens, for hundreds of years. And we keep on diverting our minds with chatter, and we instruct each other, and we console ourselves, and we have utterly forgotten them. We have so entirely forgotten them, that others have undertaken to instruct them, and we have not even perceived it. We have spoken of the division of labor with such lack of seriousness, that it is obvious that what we have said about the benefits which we have conferred on the people was simply a shameless evasion.

CHAPTER IV

.Science and art have arrogated to themselves the right of idleness, and of the enjoyment of the labor of others, and have betrayed their calling. And their errors have arisen merely because their servants, having set forth a falsely conceived principle of the division of labor, have recognized their own right to make use of the labor of others, and have lost the significance of their vocation; having taken for their aim, not the profit of the people, but the mysterious profit of science and art, and delivered themselves over to idleness and vice--not so much of the senses as of the mind.

They say, "Science and art have bestowed a great deal on mankind."

Science and art have bestowed a great deal on mankind, not because the men of art and science, under the pretext of a division of labor, live on other people, but in spite of this.

The Roman Republic was powerful, not because her citizens had the power to live a vicious life, but because among their number there were heroic citizens. It is the same with art and science. Art and science have bestowed much on mankind, but not because their followers formerly possessed on rare occasions (and now possess on every occasion) the possibility of getting rid of labor; but because there have been men of genius, who, without making use of these rights, have led mankind forward.

The class of learned men and artists, which has advanced, on the fictitious basis of a division of labor, its demands to the right of using the labors of others, cannot co-operate in the success of true science and true art, because a lie cannot bring forth the truth.

We have become so accustomed to these, our tenderly reared or weakened representatives of mental labor, that it seems to us horrible that a man of science or an artist should plough or cart manure. It seems to us that every thing would go to destruction, and that all his wisdom would be rattled out of him in the cart, and that all those grand picturesque images which he bears about in his breast would be soiled in the manure; but we have become so inured to this, that it does not strike us as strange that our servitor of science--that is to say, the servant and teacher of the truth--by making other people do for him that which he might do for himself, passes half his time in dainty eating, in smoking, in talking, in free and easy gossip, in reading the newspapers and romances, and in visiting the theatres. It is not strange to us to see our philosopher in the tavern, in the theatre, and at the ball. It is not strange in our eyes to learn that those artists who sweeten and ennoble our souls have passed their lives in drunkenness, cards, and women, if not in something worse.

Art and science are very beautiful things; but just because they are so beautiful they should not be spoiled by the compulsory combination with them of vice: that is to say, a man should not get rid of his obligation to serve his own life and that of other people by his own labor. Art and science have caused mankind to progress. Yes; but not because men of art and science, under the guise of division of labor, have rid themselves of the very first and most indisputable of human obligations,--to labor with their hands in the universal struggle of mankind with nature.

"But only the division of labor, the freedom of men of science and of art from the necessity of earning them living, has rendered possible that remarkable success of science which we behold in our day," is the answer to this. "If all were forced to till the soil, those VAST results would not have been attained which have been attained in our day; there would have been none of those STRIKING successes which have so greatly augmented man's power over nature, were it not for these astronomical discoveries WHICH ARE SO ASTOUNDING TO THE MIND OF MAN, and which have added to the security of navigation; there would be no steamers, no railways, none of those WONDERFUL bridges, tunnels, steam-engines and telegraphs, photography, telephones, sewing-machines, phonographs, electricity, telescopes, spectroscopes, microscopes, chloroform, Lister's bandages, and carbolic acid."

I will not enumerate every thing on which our age thus prides itself. This enumeration and pride of enthusiasm over ourselves and our exploits can be found in almost any newspaper and popular pamphlet. This enthusiasm over ourselves is often repeated to such a degree that none of us can sufficiently rejoice over ourselves, that we are seriously convinced that art and science have never made such progress as in our own time. And, as we are indebted for all this marvellous progress to the division of labor, why not acknowledge it?

Let us admit that the progress made in our day is noteworthy, marvellous, unusual; let us admit that we are fortunate mortals to live in such a remarkable epoch: but let us endeavor to appraise this progress, not on the basis of our self-satisfaction, but of that principle which defends itself with this progress,--the division of labor. All this progress is very amazing; but by a peculiarly unlucky chance, admitted even by the men of science, this progress has not so far improved, but it has rather rendered worse, the position of the majority, that is to say, of the workingman.

If the workingman can travel on the railway, instead of walking, still that same railway has burned down his forest, has carried off his grain under his very nose, and has brought his condition very near to slavery--to the capitalist. If, thanks to steam-engines and machines, the workingman can purchase inferior calico at a cheap rate, on the other hand these engines and machines have deprived him of work at home, and have brought him into a state of abject slavery to the manufacturer. If there are telephones and telescopes, poems, romances, theatres, ballets, symphonies, operas, picture-galleries, and so forth, on the other hand the life of the workingman has not been bettered by all this; for all of them, by the same unlucky chance, are inaccessible to him.

So that, on the whole (and even men of science admit this), up to the present time, all these remarkable discoveries and products of science and art have certainly not ameliorated the condition of the workingman, if, indeed, they have not made it worse. So that, if we set against the question as to the reality of the progress attained by the arts and sciences, not our own rapture, but that standard upon the basis of which the division of labor is defended,--the good of the laboring man,--we shall see that we have no firm foundations for that self-satisfaction in which we are so fond of indulging.

The peasant travels on the railway, the woman buys calico, in the isba (cottage) there will be a lamp instead of a pine-knot, and the peasant will light his pipe with a match,--this is convenient; but what right have I to say that the railway and the factory have proved advantageous to the people?

If the peasant rides on the railway, and buys calico, a lamp, and matches, it is only because it is impossible to forbid the peasant's buying them; but surely we are all aware that the construction of railways and factories has never been carried out for the benefit of the lower classes: so why should a casual convenience which the workingman enjoys lead to a proof of the utility of all these institutions for the people?

There is something useful in every injurious thing. After a conflagration, one can warm one's self, and light one's pipe with a firebrand; but why declare that the conflagration is beneficial?

Men of art and science might say that their pursuits are beneficial to the people, only when men of art and science have assigned to themselves the object of serving the people, as they now assign themselves the object of serving the authorities and the capitalists. We might say this if men of art and science had taken as their aim the needs of the people; but there are none such. All scientists are busy with their priestly avocations, out of which proceed investigations into protoplasm, the spectral analyses of stars, and so on. But science has never once thought of what axe or what hatchet is the most profitable to chop with, what saw is the most handy, what is the best way to mix bread, from what flour, how to set it, how to build and heat an oven, what food and drink, and what utensils, are the most convenient and advantageous under certain conditions, what mushrooms may be eaten, how to propagate them, and how to prepare them in the most suitable manner. And yet all this is the province of science.

I am aware, that, according to its own definition, science ought to be useless, i.e., science for the sake of science; but surely this is an obvious evasion. The province of science is to serve the people. We have invented telegraphs, telephones, phonographs; but what advances have we effected in the life, in the labor, of the people? We have reckoned up two millions of beetles! And we have not tamed a single animal since biblical times, when all our animals were already domesticated; but the reindeer, the stag, the partridge, the heath-cock, all remain wild.

Our botanists have discovered the cell, and in the cell protoplasm, and in that protoplasm still something more, and in that atom yet another thing. It is evident that these occupations will not end for a long time to come, because it is obvious that there can be no end to them, and therefore the scientist has no time to devote to those things which are necessary to the people. And therefore, again, from the time of Egyptian and Hebrew antiquity, when wheat and lentils had already been cultivated, down to our own times, not a single plant has been added to the food of the people, with the exception of the potato, and that was not obtained by science.

Torpedoes have been invented, and apparatus for taxation, and so forth. But the spinning-whined, the woman's weaving-loom, the plough, the hatchet, the chain, the rake, the bucket, the well- sweep, are exactly the same as they were in the days of Rurik; and if there has been any change, then that change has not been effected by scientific people.

And it is the same with the arts. We have elevated a lot of people to the rank of great writers; we have picked these writers to pieces, and have written mountains of criticism, and criticism on the critics, and criticism on the critics of the critics. And we have collected picture-galleries, and have studied different schools of art in detail; and we have so many symphonies and orchestras and operas, that it is becoming difficult even for us to listen to them. But what have we added to the popular bylini [the epic songs], legends, tales, songs? What music, what pictures, have we given to the people?

On the Nikolskaya books are manufactured for the people, and harmonicas in Tula; and in neither have we taken any part. The falsity of the whole direction of our arts and sciences is more striking and more apparent in precisely those very branches, which, it would seem, should, from their very nature, be of use to the people, and which, in consequence of their false attitude, seem rather injurious than useful. The technologist, the physician, the teacher, the artist, the author, should, in virtue of their very callings, it would seem, serve the people. And, what then? Under the present regime, they can do nothing but harm to the people.

The technologist or the mechanic has to work with capital. Without capital he is good for nothing. All his acquirements are such that for their display he requires capital, and the exploitation of the laboring-man on the largest scale; and--not to mention that he is trained to live, at the lowest, on from fifteen hundred to two thousand a year, and that, therefore, he cannot go to the country, where no one can give him such wages,--he is, by virtue of his very occupation, unfitted for serving the people. He knows how to calculate the highest mathematical arch of a bridge, how to calculate the force and transfer of the motive power, and so on; but he is confounded by the simplest questions of a peasant: how to improve a plough or a cart, or how to make irrigating canals. All this in the conditions of life in which the laboring man finds himself. Of this, he neither knows nor understands any thing,-- less, indeed, than the very stupidest peasant. Give him workshops, all sorts of workmen at his desire, an order for a machine from abroad, and he will get along. But how to devise means of lightening toil, under the conditions of labor of millions of men,-- this is what he does not and can not know; and because of his knowledge, his habits, and his demands on life, he is unfitted for this business.

In a still worse predicament is the physician. His fancied science is all so arranged, that he only knows how to heal those persons who do nothing. He requires an incalculable quantity of expensive preparations, instruments, drugs, and hygienic apparatus.

He has studied with celebrities in the capitals, who only retain patients who can be cured in the hospital, or who, in the course of their cure, can purchase the appliances requisite for healing, and even go at once from the North to the South, to some baths or other. Science is of such a nature, that every rural physic-man laments because there are no means of curing working-men, because he is so poor that he has not the means to place the sick man in the proper hygienic conditions; and at the same time this physician complains that there are no hospitals, and that he cannot get through with his work, that he needs assistants, more doctors and practitioners.

What is the inference? This: that the people's principal lack, from which diseases arise, and spread abroad, and refuse to be healed, is the lack of means of subsistence. And here Science, under the banner of the division of labor, summons her warriors to the aid of the people. Science is entirely arranged for the wealthy classes, and it has adopted for its task the healing of the people who can obtain every thing for themselves; and it attempts to heal those who possess no superfluity, by the same means.

But there are no means, and therefore it is necessary to take them from the people who are ailing, and pest-stricken, and who cannot recover for lack of means. And now the defenders of medicine for the people say that this matter has been, as yet, but little developed. Evidently it has been but little developed, because if (which God forbid!) it had been developed, and that through oppressing the people,--instead of two doctors, midwives, and practitioners in a district, twenty would have settled down, since they desire this, and half the people would have died through the difficulty of supporting this medical staff, and soon there would be no one to heal.

Scientific co-operation with the people, of which the defenders of science talk, must be something quite different. And this co- operation which should exist has not yet begun. It will begin when the man of science, technologist or physician, will not consider it legal to take from people--I will not say a hundred thousand, but even a modest ten thousand, or five hundred rubles for assisting them; but when he will live among the toiling people, under the same conditions, and exactly as they do, then he will be able to apply his knowledge to the questions of mechanics, technics, hygiene, and the healing of the laboring people. But now science, supporting itself at the expense of the working-people, has entirely forgotten the conditions of life among these people, ignores (as it puts it) these conditions, and takes very grave offence because its fancied knowledge finds no adherents among the people.

The domain of medicine, like the domain of technical science, still lies untouched. All questions as to how the time of labor is best divided, what is the best method of nourishment, with what, in what shape, and when it is best to clothe one's self, to shoe one's self, to counteract dampness and cold, how best to wash one's self, to feed the children, to swaddle them, and so on, in just those conditions in which the working-people find themselves,-- all these questions have not yet been propounded.

The same is the case with the activity of the teachers of science,-- pedagogical teachers. Exactly in the same manner science has so arranged this matter, that only wealthy people are able to study science, and teachers, like technologists and physicians, cling to money.

And this cannot be otherwise, because a school built on a model plan (as a general rule, the more scientifically built the school, the more costly it is), with pivot chains, and globes, and maps, and library, and petty text-books for teachers and scholars and pedagogues, is a sort of thing for which it would be necessary to double the taxes in every village. This science demands. The people need money for their work; and the more there is needed, the poorer they are.

Defenders of science say: "Pedagogy is even now proving of advantage to the people, but give it a chance to develop, and then it will do still better." Yes, if it does develop, and instead of twenty schools in a district there are a hundred, and all scientific, and if the people support these schools, they will grow poorer than ever, and they will more than ever need work for their children's sake. "What is to be done?" they say to this. The government will build the schools, and will make education obligatory, as it is in Europe; but again, surely, the money is taken from the people just the same, and it will be harder to work, and they will have less leisure for work, and there will be no education even by compulsion. Again the sole salvation is this: that the teacher should live under the conditions of the working- men, and should teach for that compensation which they give him freely and voluntarily.

Such is the false course of science, which deprives it of the power of fulfilling its obligation, which is, to serve the people.

But in nothing is this false course of science so obviously apparent, as in the vocation of art, which, from its very significance, ought to be accessible to the people. Science may fall back on its stupid excuse, that science acts for science, and that when it turns out learned men it is laboring for the people; but art, if it is art, should be accessible to all the people, and in particular to those in whose name it is executed. And our definition of art, in a striking manner, convicts those who busy themselves with art, of their lack of desire, lack of knowledge, and lack of power, to be useful to the people.

The painter, for the production of his great works, must have a studio of at least such dimensions that a whole association of carpenters (forty in number) or shoemakers, now sickening or stifling in lairs, would be able to work in it. But this is not all; he must have a model, costumes, travels. Millions are expended on the encouragement of art, and the products of this art are both incomprehensible and useless to the people. Musicians, in order to express their grand ideas, must assemble two hundred men in white neckties, or in costumes, and spend hundreds of thousands of rubles for the equipment of an opera. And the products of this art cannot evoke from the people--even if the latter could at any time enjoy it--any thing except amazement and ennui.

Writers--authors--it appears, do not require surroundings, studios, models, orchestras, and actors; but it then appears that the author needs (not to mention comfort in his quarters) all the dainties of life for the preparation of his great works, travels, palaces, cabinets, libraries, the pleasures of art, visits to theatres, concerts, the baths, and so on. If he does not earn a fortune for himself, he is granted a pension, in order that he may compose the better. And again, these compositions, so prized by us, remain useless lumber for the people, and utterly unserviceable to them.

And if still more of these dealers in spiritual nourishment are developed further, as men of science desire, and a studio is erected in every village; if an orchestra is set up, and authors are supported in those conditions which artistic people regard as indispensable for themselves,--I imagine that the working-classes will sooner take an oath never to look at any pictures, never to listen to a symphony, never to read poetry or novels, than to feed all these persons.

And why, apparently, should art not be of service to the people? In every cottage there are images and pictures; every peasant man and woman sings; many own harmonicas; and all recite stories and verses, and many read. It is as if those two things which are made for each other--the lock and the key-- had parted company; they have sprung so far apart, that not even the possibility of uniting them presents itself. Tell the artist that he should paint without a studio, model, or costumes, and that he should paint five-kopek pictures, and he will say that that is tantamount to abandoning his art, as he understands it. Tell the musician that he should play on the harmonica, and teach the women to sing songs; say to the poet, to the author, that he ought to cast aside his poems and romances, and compose song-books, tales, and stories, comprehensible to the uneducated people,--they will say that you are mad.

The service of the people by science and art will only be performed when people, dwelling in the midst of the common folk, and, like the common folk, putting forward no demands, claiming no rights, shall offer to the common folk their scientific and artistic services; the acceptance or rejection of which shall depend wholly on the will of the common folk.

It is said that the activity of science and art has aided in the forward march of mankind,--meaning by this activity, that which is now called by that name; which is the same as saying that an unskilled banging of oars on a vessel that is floating with the tide, which merely hinders the progress of the vessel, is assisting the movement of the ship. It only retards it. The so-called division of labor, which has become in our day the condition of activity of men of science and art, was, and has remained, the chief cause of the tardy forward movement of mankind.

The proofs of this lie in that confession of all men of science, that the gains of science and art are inaccessible to the laboring masses, in consequence of the faulty distribution of riches. The irregularity of this distribution does not decrease in proportion to the progress of science and art, but only increases. Men of art and science assume an air of deep pity for this unfortunate circumstance which does not depend upon them. But this unfortunate circumstance is produced by themselves; for this irregular distribution of wealth flows solely from the theory of the division of labor.

Science maintains the division of labor as a unalterable law; it sees that the distribution of wealth, founded on the division of labor, is wrong and ruinous; and it affirms that its activity, which recognizes the division of labor, will lead people to bliss. The result is, that some people make use of the labor of others; but that, if they shall make use of the labor of others for a very long period of time, and in still larger measure, then this wrongful distribution of wealth, i.e., the use of the labor of others, will come to an end.

Men stand beside a constantly swelling spring of water, and are occupied with the problem of diverting it to one side, away from the thirsty people, and they assert that they are producing this water, and that soon enough will be collected for all. But this water which has flowed, and which still flows unceasingly, and nourishes all mankind, not only is not the result of the activity of the men who, standing at its source, turn it aside, but this water flows and gushes out, in spite of the efforts of these men to obstruct its flow.

There have always existed a true science, and a true art; but true science and art are not such because they called themselves by that name. It always seems to those who claim at any given period to be the representatives of science and art, that they have performed, and are performing, and--most of all--that they will presently perform, the most amazing marvels, and that beside them there never has been and there is not any science or any art. Thus it seemed to the sophists, the scholastics, the alchemists, the cabalists, the talmudists; and thus it seems to our own scientific science, and to our art for the sake of art.

CHAPTER V

."But art,--science! You repudiate art and science; that is, you repudiate that by which mankind lives!" People are constantly making this--it is not a reply--to me, and they employ this mode of reception in order to reject my deductions without examining into them. "He repudiates science and art, he wants to send people back again into a savage state; so what is the use of listening to him and of talking to him?" But this is unjust. I not only do not repudiate art and science, but, in the name of that which is true art and true science, I say that which I do say; merely in order that mankind may emerge from that savage state into which it will speedily fall, thanks to the erroneous teaching of our time,--only for this purpose do I say that which I say.

Art and science are as indispensable as food and drink and clothing,--more indispensable even; but they become so, not because we decide that what we designate as art and science are indispensable, but simply because they really are indispensable to people.

Surely, if hay is prepared for the bodily nourishment of men, the fact that we are convinced that hay is the proper food for man will not make hay the food of man. Surely I cannot say, "Why do not you eat hay, when it is the indispensable food?" Food is indispensable, but it may happen that that which I offer is not food at all. This same thing has occurred with our art and science. It seems to us, that if we add to a Greek word the word "logy," and call that a science, it will be a science; and, if we call any abominable thing- -like the dancing of nude females--by a Greek word, choreography, that that is art, and that it will be art. But no matter how much we may say this, the business with which we occupy ourselves when we count beetles, and investigate the chemical constituents of the stars in the Milky Way, when we paint nymphs and compose novels and symphonies,--our business will not become either art or science until such time as it is accepted by those people for whom it is wrought.

If it were decided that only certain people should produce food, and if all the rest were forbidden to do this, or if they were rendered incapable of producing food, I suppose that the quality of food would be lowered. If the people who enjoyed the monopoly of producing food were Russian peasants, there would be no other food than black bread and cabbage-soup, and so on, and kvas,--nothing except what they like, and what is agreeable to them. The same thing would happen in the case of that loftiest human pursuit, of arts and sciences, if one caste were to arrogate to itself a monopoly of them: but with this sole difference, that, in the matter of bodily food, there can be no great departure from nature, and bread and cabbage-soup, although not very savory viands, are fit for consumption; but in spiritual food, there may exist the very greatest departures from nature, and some people may feed themselves for a long time on poisonous spiritual nourishment, which is directly unsuitable for, or injurious to, them; they may slowly kill themselves with spiritual opium or liquors, and they may offer this same food to the masses.

It is this very thing that is going on among us. And it has come about because the position of men of science and art is a privileged one, because art and science (in our day), in our world, are not at all a rational occupation of all mankind without exception, exerting their best powers for the service of art and science, but an occupation of a restricted circle of people holding a monopoly of these industries, and entitling themselves men of art and science, and who have, therefore, perverted the very idea of art and science, and have lost all the meaning of their vocation, and who are only concerned in amusing and rescuing from crushing ennui their tiny circle of idle mouths.

Ever since men have existed, they have always had science and art in the simplest and broadest sense of the term. Science, in the sense of the whole of knowledge acquired by mankind, exists and always has existed, and life without it is not conceivable; and there is no possibility of either attacking or defending science, taken in this sense.

But the point lies here,--that the scope of the knowledge of all mankind as a whole is so multifarious, ranging from the knowledge of how to extract iron to the knowledge of the movements of the planets, that man loses himself in this multitude of existing knowledge,--knowledge capable of ENDLESS possibilities, if he have no guiding thread, by the aid of which he can classify this knowledge, and arrange the branches according to the degrees of their significance and importance.

Before a man undertakes to learn any thing whatever, he must make up his mind that that branch of knowledge is of weight to him, and of more weight and importance than the countless other objects of study with which he is surrounded. Before undertaking the study of any thing, a man decides for what purpose he is studying this subject, and not the others. But to study every thing, as the men of scientific science in our day preach, without any idea of what is to come out of such study, is downright impossible, because the number of subjects of study is ENDLESS; and hence, no matter how many branches we may acquire, their acquisition can possess no significance or reason. And, therefore, in ancient times, down to even a very recent date, until the appearance of scientific science, man's highest wisdom consisted in finding that guiding thread, according to which the knowledge of men should be classified as being of primary or of secondary importance. And this knowledge, which forms the guide to all other branches of knowledge, men have always called science in the strictest acceptation of the word. And such science there has always been, even down to our own day, in all human communities which have emerged from their primal state of savagery.

Ever since mankind has existed, teachers have always arisen among peoples, who have enunciated science in this restricted sense,--the science of what it is most useful for man to know. This science has always had for its object the knowledge of what is the true ground of the well-being of each individual man, and of all men, and why. Such was the science of Confucius, of Buddha, of Socrates, of Mahomet, and of others; such is this science as they understood it, and as all men--with the exception of our little circle of so-called cultured people--understand it. This science has not only always occupied the highest place, but has been the only and sole science, from which the standing of the rest has been determined. And this was the case, not in the least because, as the so-called scientific people of our day think, cunning priestly teachers of this science attributed to it such significance, but because in reality, as every one knows, both by personal experience and by reflection, there can be no science except the science of that in which the destiny and welfare of man consist. For the objects of science are INCALCULABLE in number,--I undermine the word "incalculable" in the exact sense in which I understand it,--and without the knowledge of that in which the destiny and welfare of all men consist, there is no possibility of making a choice amid this interminable multitude of subjects; and therefore, without this knowledge, all other arts and branches of learning will become, as they have become among us, an idle and hurtful diversion.

Mankind has existed and existed, and never has it existed without the science of that in which the destiny and the welfare of men consist. It is true that the science of the welfare of men appears different on superficial observation, among the Buddhists, the Brahmins, the Hebrews, the Confucians, the Tauists; but nevertheless, wherever we hear of men who have emerged from a state of savagery, we find this science. And all of a sudden it appears that the men of our day have decided that this same science, which has hitherto served as the guiding thread of all human knowledge, is the very thing which hinders every thing. Men erect buildings; and one architect has made one estimate of cost, a second has made another, and a third yet another. The estimates differ somewhat; but they are correct, so that any one can see, that, if the whole is carried out in accordance with the calculations, the building will be erected. Along come people, and assert that the chief point lies in having no estimates, and that it should be built thus--by the eye. And this "thus," men call the most accurate of scientific science. Men repudiate every science, the very substance of science,--the definition of the destiny and the welfare of men,--and this repudiation they designate as science.

Ever since men have existed, great minds have been born into their midst, which, in the conflict with reason and conscience, have put to themselves questions as to "what constitutes welfare,--the destiny and welfare, not of myself alone, but of every man?" What does that power which has created and which leads me, demand of me and of every man? And what is it necessary for me to do, in order to comply with the requirements imposed upon me by the demands of individual and universal welfare? They have asked themselves: "I am a whole, and also a part of something infinite, eternal; what, then, are my relations to other parts similar to myself, to men and to the whole--to the world?"

And from the voices of conscience and of reason, and from a comparison of what their contemporaries and men who had lived before them, and who had propounded to themselves the same questions, had said, these great teachers have deduced their doctrines, which were simple, clear, intelligible to all men, and always such as were susceptible of fulfilment. Such men have existed of the first, second, third, and lowest ranks. The world is full of such men. Every living man propounds the question to himself, how to reconcile the demands of welfare, and of his personal existence, with conscience and reason; and from this universal labor, slowly but uninterruptedly, new forms of life, which are more in accord with the requirements of reason and of conscience, are worked out.

All at once, a new caste of people makes its appearance, and they say, "All this is nonsense; all this must be abandoned." This is the deductive method of ratiocination (wherein lies the difference between the deductive and the inductive method, no one can understand); these are the dogmas of the technological and metaphysical period. Every thing that these men discover by inward experience, and which they communicate to one another, concerning their knowledge of the law of their existence (of their functional activity, according to their own jargon), every thing that the grandest minds of mankind have accomplished in this direction, since the beginning of the world,--all this is nonsense, and has no weight whatever. According to this new doctrine, it appears that you are cells: and that you, as a cell, have a very definite functional activity, which you not only fulfil, but which you infallibly feel within you; and that you are a thinking, talking, understanding cell, and that you, for this reason, can ask another similar talking cell whether it is just the same, and in this way verify your own experience; that you can take advantage of the fact that speaking cells, which have lived before you, have written on the same subject, and that you have millions of cells which confirm your observations by their agreement with the cells which have written down their thoughts,--all this signifies nothing; all this is an evil and an erroneous method.

The true scientific method is this: If you wish to know in what the destiny and the welfare of all mankind and of all the world consists, you must, first of all, cease to listen to the voices of your conscience and of your reason, which present themselves in you and in others like you; you must cease to believe all that the great teachers of mankind have said with regard to your conscience and reason, and you must consider all this as nonsense, and begin all over again. And, in order to understand every thing from the beginning, you must look through microscopes at the movements of amoebae, and cells in worms, or, with still greater composure, believe in every thing that men with a diploma of infallibility shall say to you about them. And as you gaze at the movements of these cells, or read about what others have seen, you must attribute to these cells your own human sensations and calculations as to what they desire, whither they are directing themselves, how they compare and discuss, and to what they have become accustomed; and from these observations (in which there is not a word about an error of thought or of expression) you must deduce a conclusion by analogy as to what you are, what is your destiny, wherein lies the welfare of yourself and of other cells like you. In order to understand yourself, you must study not only the worms which you see, but microscopic creatures which you can barely see, and transformations from one set of creatures into others, which no one has ever beheld, and which you, most assuredly, will never behold. And the same with art. Where there has been true science, art has always been its exponent.

Ever since men have been in existence, they have been in the habit of deducing, from all pursuits, the expressions of various branches of learning concerning the destiny and the welfare of man, and the expression of this knowledge has been art in the strict sense of the word.

Ever since men have existed, there have been those who were peculiarly sensitive and responsive to the doctrine regarding the destiny and welfare of man; who have given expression to their own and the popular conflict, to the delusions which lead them astray from their destinies, their sufferings in this conflict, their hopes in the triumph of good, them despair over the triumph of evil, and their raptures in the consciousness of the approaching bliss of man, on viol and tabret, in images and words. Always, down to the most recent times, art has served science and life,--only then was it what has been so highly esteemed of men. But art, in its capacity of an important human activity, disappeared simultaneously with the substitution for the genuine science of destiny and welfare, of the science of any thing you choose to fancy. Art has existed among all peoples, and will exist until that which among us is scornfully called religion has come to be considered the only science.

In our European world, so long as there existed a Church, as the doctrine of destiny and welfare, and so long as the Church was regarded as the only true science, art served the Church, and remained true art: but as soon as art abandoned the Church, and began to serve science, while science served whatever came to hand, art lost its significance. And notwithstanding the rights claimed on the score of ancient memories, and of the clumsy assertion which only proves its loss of its calling, that art serves art, it has become a trade, providing men with something agreeable; and as such, it inevitably comes into the category of choreographic, culinary, hair-dressing, and cosmetic arts, whose practitioners designate themselves as artists, with the same right as the poets, printers, and musicians of our day.

Glance backward into the past, and you will see that in the course of thousands of years, out of milliards of people, only half a score of

Confucius', Buddhas, Solomons, Socrates, Solons, and Homers have been produced. Evidently, they are rarely met with among men, in spite of the fact that these men have not been selected from a single caste, but from mankind at large. Evidently, these true teachers and artists and learned men, the purveyors of spiritual nourishment, are rare. And it is not without reason that mankind has valued and still values them so highly.

But it now appears, that all these great factors in the science and art of the past are no longer of use to us. Nowadays, scientific and artistic authorities can, in accordance with the law of division of labor, be turned out by factory methods; and, in one decade, more great men have been manufactured in art and science, than have ever been born of such among all nations, since the foundation of the world. Nowadays there is a guild of learned men and artists, and they prepare, by perfected methods, all that spiritual food which man requires. And they have prepared so much of it, that it is no longer necessary to refer to the elder authorities, who have preceded them,--not only to the ancients, but to those much nearer to us. All that was the activity of the theological and metaphysical period,--all that must be wiped out: but the true, the rational activity began, say, fifty years ago, and in the course of those fifty years we have made so many great men, that there are about ten great men to every branch of science. And there have come to be so many sciences, that, fortunately, it is easy to make them. All that is required is to add the Greek word "logy" to the name, and force them to conform to a set rubric, and the science is all complete. They have created so many sciences, that not only can no one man know them all, but not a single individual can remember all the titles of all the existing sciences; the titles alone form a thick lexicon, and new sciences are manufactured every day. They have been manufactured on the pattern of that Finnish teacher who taught the landed proprietor's children Finnish instead of French. Every thing has been excellently inculcated; but there is one objection,--that no one except ourselves can understand any thing of it, and all this is reckoned as utterly useless nonsense. However, there is an explanation even for this. People do not appreciate the full value of scientific science, because they are under the influence of the theological period, that profound period when all the people, both among the Hebrews, and the Chinese, and the Indians, and the Greeks, understood every thing that their great teachers said to them.

But, from whatever cause this has come about, the fact remains, that sciences and arts have always existed among mankind, and, when they really did exist, they were useful and intelligible to all the people. But we practise something which we call science and art, but it appears that what we do is unnecessary and unintelligible to man. And hence, however beautiful may be the things that we accomplish, we have no right to call them arts and sciences.

CHAPTER VI

."But you only furnish a different definition of arts and sciences, which is stricter, and is incompatible with science," I shall be told in answer to this; "nevertheless, scientific and artistic activity does still exist. There are the Galileos, Brunos, Homers, Michael Angelos, Beethovens, and all the lesser learned men and artists, who have consecrated their entire lives to the service of science and art, and who were, and will remain, the benefactors of mankind."

Generally this is what people say, striving to forget that new principle of the division of labor, on the basis of which science and art now occupy their privileged position, and on whose basis we are now enabled to decide without grounds, but by a given standard: Is there, or is there not, any foundation for that activity which calls itself science and art, to so magnify itself?

When the Egyptian or the Grecian priests produced their mysteries, which were unintelligible to any one, and stated concerning these mysteries that all science and all art were contained in them, I could not verify the reality of their science on the basis of the benefit procured by them to the people, because science, according to their assertions, was supernatural. But now we all possess a very simple and clear definition of the activity of art and science, which excludes every thing supernatural: science and art promise to carry out the mental activity of mankind, for the welfare of society, or of all the human race.

The definition of scientific science and art is entirely correct; but, unfortunately, the activity of the present arts and sciences does not come under this head. Some of them are directly injurious, others are useless, others still are worthless,--good only for the wealthy. They do not fulfil that which, by their own definition, they have undertaken to accomplish; and hence they have as little right to regard themselves as men of art and science, as a corrupt priesthood, which does not fulfil the obligations which it has assumed, has the right to regard itself as the bearer of divine truth.

And it can be understood why the makers of the present arts and sciences have not fulfilled, and cannot fulfil, their vocation. They do not fulfil it, because out of their obligations they have erected a right.

Scientific and artistic activity, in its real sense, is only fruitful when it knows no rights, but recognizes only obligations. Only because it is its property to be always thus, does mankind so highly prize this activity. If men really were called to the service of others through artistic work, they would see in that work only obligation, and they would fulfil it with toil, with privations, and with self-abnegation.

The thinker or the artist will never sit calmly on Olympian heights, as we have become accustomed to represent them to ourselves. The thinker or the artist should suffer in company with the people, in order that he may find salvation or consolation. Besides this, he will suffer because he is always and eternally in turmoil and agitation: he might decide and say that that which would confer welfare on men, would free them from suffering, would afford them consolation; but he has not said so, and has not presented it as he should have done; he has not decided, and he has not spoken; and to- morrow, possibly, it will be too late,--he will die. And therefore suffering and self-sacrifice will always be the lot of the thinker and the artist.

Not of this description will be the thinker and artist who is reared in an establishment where, apparently, they manufacture the learned man or the artist (but in point of fact, they manufacture destroyers of science and of art), who receives a diploma and a certificate, who would be glad not to think and not to express that which is imposed on his soul, but who cannot avoid doing that to which two irresistible forces draw him,--an inward prompting, and the demand of men.

There will be no sleek, plump, self-satisfied thinkers and artists. Spiritual activity, and its expression, which are actually necessary to others, are the most burdensome of all man's avocations; a cross, as the Gospels phrase it. And the sole indubitable sign of the presence of a vocation is self-devotion, the sacrifice of self for the manifestation of the power that is imposed upon man for the benefit of others.

It is possible to study out how many beetles there are in the world, to view the spots on the sun, to write romances and operas, without suffering; but it is impossible, without self-sacrifice, to instruct people in their true happiness, which consists solely in renunciation of self and the service of others, and to give strong expression to this doctrine, without self-sacrifice.

Christ did not die on the cross in vain; not in vain does the sacrifice of suffering conquer all things.

But our art and science are provided with certificates and diplomas; and the only anxiety of all men is, how to still better guarantee them, i.e., how to render the service of the people impracticable for them.

True art and true science possess two unmistakable marks: the first, an inward mark, which is this, that the servitor of art and science will fulfil his vocation, not for profit but with self- sacrifice; and the second, an external sign,--his productions will be intelligible to all the people whose welfare he has in view.

No matter what people have fixed upon as their vocation and their welfare, science will be the doctrine of this vocation and welfare, and art will be the expression of that doctrine. That which is called science and art, among us, is the product of idle minds and feelings, which have for their object to tickle similar idle minds and feelings. Our arts and sciences are incomprehensible, and say nothing to the people, for they have not the welfare of the common people in view.

Ever since the life of men has been known to us, we find, always and everywhere, the reigning doctrine falsely designating itself as science, not manifesting itself to the common people, but obscuring for them the meaning of life. Thus it was among the Greeks the sophists, then among the Christians the mystics, gnostics, scholastics, among the Hebrews the Talmudists and Cabalists, and so on everywhere, down to our own times.

How fortunate it is for us that we live in so peculiar an age, when that mental activity which calls itself science, not only does not err, but finds itself, as we are assured, in a remarkably flourishing condition! Does not this peculiar good fortune arise from the fact that man can not and will not see his own hideousness? Why is there nothing left of those sciences, and sophists, and Cabalists, and Talmudists, but words, while we are so exceptionally happy? Surely the signs are identical. There is the same self-satisfaction and blind confidence that we, precisely we, and only we, are on the right path, and that the real thing is only beginning with us. There is the same expectation that we shall discover something remarkable; and that chief sign which leads us astray convicts us of our error: all our wisdom remains with us, and the common people do not understand, and do not accept, and do not need it.

Our position is a very difficult one, but why not look at it squarely?

It is time to recover our senses, and to scrutinize ourselves. Surely we are nothing else than the scribes and Pharisees, who sit in Moses' seat, and who have taken the keys of the kingdom of heaven, and will neither go in ourselves, nor permit others to go in. Surely we, the high priests of science and art, are ourselves worthless deceivers, possessing much less right to our position than the most crafty and depraved priests. Surely we have no justification for our privileged position. The priests had a right to their position: they declared that they taught the people life and salvation. But we have taken their place, and we do not instruct the people in life,--we even admit that such instruction is unnecessary,--but we educate our children in the same Talmudic-Greek and Latin grammar, in order that they may be able to pursue the same life of parasites which we lead ourselves. We say, "There used to be castes, but there are none among us." But what does it mean, that some people and their children toil, while other people and their children do not toil?

Bring hither an Indian ignorant of our language, and show him European life, and our life, for several generations, and he will recognize the same leading, well-defined castes--of laborers and non-laborers--as there are in his own country. And as in his land, so in ours, the right of refusing to labor is conferred by a peculiar consecration, which we call science and art, or, in general terms, culture. It is this culture, and all the distortions of sense connected with it, which have brought us to that marvellous madness, in consequence of which we do not see that which is so clear and indubitable.

CHAPTER VII

.Then, what is to be done? What are we to do?

This question, which includes within itself both an admission that our life is evil and wrong, and in connection with this,--as though it were an exercise for it,--that it is impossible, nevertheless, to change it, this question I have heard, and I continue to hear, on all sides. I have described my own sufferings, my own gropings, and my own solution of this question. I am the same kind of a man as everybody else; and if I am in any wise distinguished from the average man of our circle, it is chiefly in this respect, that I, more than the average man, have served and winked at the false doctrine of our world; I have received more approbation from men professing the prevailing doctrine: and therefore, more than others, have I become depraved, and wandered from the path. And therefore I think that the solution of the problem, which I have found in my own case, will be applicable to all sincere people who are propounding the same question to themselves.

First of all, in answer to the question, "What is to be done?" I told myself: "I must lie neither to other people nor to myself. I must not fear the truth, whithersoever it may lead me."

We all know what it means to lie to other people, but we are not afraid to lie to ourselves; yet the very worst downright lie, to other people, is not to be compared in its consequences with the lie to ourselves, upon which we base our whole life.

This is the lie of which we must not be guilty if we are to be in a position to answer the question: "What is to be done?" And, in fact, how am I to answer the question, "What is to be done?" when every thing that I do, when my whole life, is founded on a lie, and when I carefully parade this lie as the truth before others and before myself? Not to lie, in this sense, means not to fear the truth, not to devise subterfuges, and not to accept the subterfuges devised by others for the purpose of hiding from myself the deductions of my reason and my conscience; not to fear to part company with all those who surround me, and to remain alone in company with reason and conscience; not to fear that position to which the truth shall lead me, being firmly convinced that that position to which truth and conscience shall conduct me, however singular it may be, cannot be worse than the one which is founded on a lie. Not to lie, in our position of privileged persons of mental labor, means, not to be afraid to reckon one's self up wrongly. It is possible that you are already so deeply indebted that you cannot take stock of yourself; but to whatever extent this may be the case, however long may be the account, however far you have strayed from the path, it is still better than to continue therein. A lie to other people is not alone unprofitable; every matter is settled more directly and more speedily by the truth than by a lie. A lie to others only entangles matters, and delays the settlement; but a lie to one's self, set forth as the truth, ruins a man's whole life. If a man, having entered on the wrong path, assumes that it is the true one, then every step that he takes on that path removes him farther from his goal. If a man who has long been travelling on this false path divines for himself, or is informed by some one, that his course is a mistaken one, but grows alarmed at the idea that he has wandered very far astray and tries to convince himself that he may, possibly, still strike into the right road, then he never will get into it. If a man quails before the truth, and, on perceiving it, does not accept it, but does accept a lie for the truth, then he never will learn what he ought to do. We, the not only wealthy, but privileged and so-called cultivated persons, have advanced so far on the wrong road, that a great deal of determination, or a very great deal of suffering on the wrong road, is required, in order to bring us to our senses and to the acknowledgment of the lie in which we are living. I have perceived the lie of our lives, thanks to the sufferings which the false path entailed upon me, and, having recognized the falseness of this path on which I stood, I have had the boldness to go at first in thought only--whither reason and conscience led me, without reflecting where they would bring me out. And I have been rewarded for this boldness.

All the complicated, broken, tangled, and incoherent phenomena of life surrounding me, have suddenly become clear; and my position in the midst of these phenomena, which was formerly strange and burdensome, has become, all at once, natural, and easy to bear.

In this new position, my activity was defined with perfect accuracy; not at all as it had previously presented itself to me, but as a new and much more peaceful, loving, and joyous activity. The very thing which had formerly terrified me, now began to attract me. Hence I think, that the man who will honestly put to himself the question, "What is to be done?" and, replying to this query, will not lie to himself, but will go whither his reason leads, has already solved the problem.

There is only one thing that can hinder him in his search for an issue,--an erroneously lofty idea of himself and of his position. This was the case with me; and then another, arising from the first answer to the question: "What is to be done?" consisted for me in this, that it was necessary for me to repent, in the full sense of that word,--i.e., to entirely alter my conception of my position and my activity; to confess the hurtfulness and emptiness of my activity, instead of its utility and gravity; to confess my own ignorance instead of culture; to confess my immorality and harshness in the place of my kindness and morality; instead of my elevation, to acknowledge my lowliness. I say, that in addition to not lying to myself, I had to repent, because, although the one flows from the other, a false conception of my lofty importance had so grown up with me, that, until I sincerely repented and cut myself free from that false estimate which I had formed of myself, I did not perceive the greater part of the lie of which I had been guilty to myself. Only when I had repented, that is to say, when I had ceased to look upon myself as a regular man, and had begun to regard myself as a man exactly like every one else,--only then did my path become clear before me. Before that time I had not been able to answer the question: "What is to be done?" because I had stated the question itself wrongly.

As long as I did not repent, I put the question thus: "What sphere of activity should I choose, I, the man who has received the education and the talents which have fallen to my shame? How, in this fashion, make recompense with that education and those talents, for what I have taken, and for what I still take, from the people?" This question was wrong, because it contained a false representation, to the effect that I was not a man just like them, but a peculiar man called to serve the people with those talents and with that education which I had won by the efforts of forty years.

I propounded the query to myself; but, in reality, I had answered it in advance, in that I had in advance defined the sort of activity which was agreeable to me, and by which I was called upon to serve the people. I had, in fact, asked myself: "In what manner could I, so very fine a writer, who had acquired so much learning and talents, make use of them for the benefit of the people?"

But the question should have been put as it would have stood for a learned rabbi who had gone through the course of the Talmud, and had learned by heart the number of letters in all the holy books, and all the fine points of his art. The question for me, as for the rabbi, should stand thus: "What am I, who have spent, owing to the misfortune of my surroundings, the year's best fitted for study in the acquisition of grammar, geography, judicial science, poetry, novels and romances, the French language, pianoforte playing, philosophical theories, and military exercises, instead of inuring myself to labor; what am I, who have passed the best years of my life in idle occupations which are corrupting to the soul,--what am I to do in defiance of these unfortunate conditions of the past, in order that I may requite those people who during the whole time have fed and clothed, yes, and who even now continue to feed and clothe me?" Had the question then stood as it stands before me now, after I have repented,--"What am I, so corrupt a man, to do?" the answer would have been easy: "To strive, first of all, to support myself honestly; that is, to learn not to live upon others; and while I am learning, and when I have learned this, to render aid on all possible occasions to the people, with my hands, and my feet, and my brain, and my heart, and with every thing to which the people should present a claim."

And therefore I say, that for the man of our circle, in addition to not lying to himself or to others, repentance is also necessary, and that he should scrape from himself that pride which has sprung up in us, in our culture, in our refinements, in our talents; and that he should confess that he is not a benefactor of the people and a distinguished man, who does not refuse to share with the people his useful acquirements, but that he should confess himself to be a thoroughly guilty, corrupt, and good-for-nothing man, who desires to reform himself and not to behave benevolently towards the people, but simply to cease wounding and insulting them.

I often hear the questions of good young men who sympathize with the renunciatory part of my writings, and who ask, "Well, and what then shall I do? What am I to do, now that I have finished my course in the university, or in some other institution, in order that I may be of use?" Young men ask this, and in the depths of their soul it is already decided that the education which they have received constitutes their privilege and that they desire to serve the people precisely by means of thus superiority. And hence, one thing which they will in no wise do, is to bear themselves honestly and critically towards that which they call their culture, and ask themselves, are those qualities which they call their culture good or bad? If they will do this, they will infallibly be led to see the necessity of renouncing their culture, and the necessity of beginning to learn all over again; and this is the one indispensable thing. They can in no wise solve the problem, "What to do?" because this question does not stand before them as it should stand. The question must stand thus: "In what manner am I, a helpless, useless man, who, owing to the misfortune of my conditions, have wasted my best years of study in conning the scientific Talmud which corrupts soul and body, to correct this mistake, and learn to serve the people?" But it presents itself to them thus: "How am I, a man who has acquired so much very fine learning, to turn this very fine learning to the use of the people?" And such a man will never answer the question, "What is to be done?" until he repents. And repentance is not terrible, just as truth is not terrible, and it is equally joyful and fruitful. It is only necessary to accept the truth wholly, and to repent wholly, in order to understand that no one possesses any rights, privileges, or peculiarities in the matter of this life of ours, but that there are no ends or bounds to obligation, and that a man's first and most indubitable duty is to take part in the struggle with nature for his own life and for the lives of others.

And this confession of a man's obligation constitutes the gist of the third answer to the question, "What is to be done?"

I tried not to lie to myself: I tried to cast out from myself the remains of my false conceptions of the importance of my education and talents, and to repent; but on the way to a decision of the question, "What to do?" a fresh difficulty arose. There are so many different occupations, that an indication was necessary as to the precise one which was to be adopted. And the answer to this question was furnished me by sincere repentance for the evil in which I had lived.

"What to do? Precisely what to do?" all ask, and that is what I also asked so long as, under the influence of my exalted idea of any own importance, I did not perceive that my first and unquestionable duty was to feed myself, to clothe myself, to furnish my own fuel, to do my own building, and, by so doing, to serve others, because, ever since the would has existed, the first and indubitable duty of every man has consisted and does consist in this.

In fact, no matter what a man may have assumed to be his vocation,-- whether it be to govern people, to defend his fellow-countrymen, to divine service, to instruct others, to invent means to heighten the pleasures of life, to discover the laws of the world, to incorporate eternal truths in artistic representations,--the duty of a reasonable man is to take part in the struggle with nature, for the sustenance of his own life and of that of others. This obligation is the first of all, because what people need most of all is their life; and therefore, in order to defend and instruct the people, and render their lives more agreeable, it is requisite to preserve that life itself, while my refusal to share in the struggle, my monopoly of the labors of others, is equivalent to annihilation of the lives of others. And, therefore, it is not rational to serve the lives of men by annihilating the lives of men; and it is impossible to say that I am serving men, when, by my life, I am obviously injuring them.

A man's obligation to struggle with nature for the acquisition of the means of livelihood will always be the first and most unquestionable of all obligations, because this obligation is a law of life, departure from which entails the inevitable punishment of either bodily or mental annihilation of the life of man. If a man living alone excuses himself from the obligation of struggling with nature, he is immediately punished, in that his body perishes. But if a man excuses himself from this obligation by making other people fulfil it for him, then also he is immediately punished by the annihilation of his mental life; that is to say, of the life which possesses rational thought.

In this one act, man receives--if the two things are to be separated--full satisfaction of the bodily and spiritual demands of his nature. The feeding, clothing, and taking care of himself and his family, constitute the satisfaction of the bodily demands and requirements; and doing the same for other people, constitutes the satisfaction of his spiritual requirements. Every other employment of man is only legal when it is directed to the satisfaction of this very first duty of man; for the fulfilment of this duty constitutes the whole life of man.

I had been so turned about by my previous life, this first and indubitable law of God or of nature is so concealed in our sphere of society, that the fulfilment of this law seemed to me strange, terrible, even shameful; as though the fulfilment of an eternal, unquestionable law, and not the departure from it, can be terrible, strange, and shameful.

At first it seemed to me that the fulfilment of this matter required some preparation, arrangement or community of men, holding similar views,--the consent of one's family, life in the country; it seemed to me disgraceful to make a show of myself before people, to undertake a thing so improper in our conditions of existence, as bodily toil, and I did not know how to set about it. But it was only necessary for me to understand that this is no exclusive occupation which requires to be invented and arranged for, but that this employment was merely a return from the false position in which I found myself, to a natural one; was only a rectification of that lie in which I was living. I had only to recognize this fact, and all these difficulties vanished. It was not in the least necessary to make preparations and arrangements, and to await the consent of others, for, no matter in what position I had found myself, there had always been people who had fed, clothed and warmed me, in addition to themselves; and everywhere, under all conditions, I could do the same for myself and for them, if I had the time and the strength. Neither could I experience false shame in an unwonted occupation, no matter how surprising it might be to people, because, through not doing it, I had already experienced not false but real shame.

And when I had reached this confession and the practical deduction from it, I was fully rewarded for not having quailed before the deductions of reason, and for following whither they led me. On arriving at this practical deduction, I was amazed at the ease and simplicity with which all the problems which had previously seemed to me so difficult and so complicated, were solved.

To the question, "What is it necessary to do?" the most indubitable answer presented itself: first of all, that which it was necessary for me to do was, to attend to my own samovar, my own stove, my own water, my own clothing; to every thing that I could do for myself. To the question, "Will it not seem strange to people if you do this?" it appeared that this strangeness lasted only a week, and after the lapse of that week, it would have seemed strange had I returned to my former conditions of life. With regard to the question, "Is it necessary to organize this physical labor, to institute an association in the country, on my land?" it appeared that nothing of the sort was necessary; that labor, if it does not aim at the acquisition of all possible leisure, and the enjoyment of the labor of others,--like the labor of people bent on accumulating money,--but if it have for its object the satisfaction of requirements, will itself be drawn from the city to the country, to the land, where this labor is the most fruitful and cheerful. But it is not requisite to institute any association, because the man who labors, naturally and of himself, attaches himself to the existing association of laboring men.

To the question, whether this labor would not monopolize all my time, and deprive me of those intellectual pursuits which I love, to which I am accustomed, and which, in my moments of self-conceit, I regard as not useless to others? I received a most unexpected reply. The energy of my intellectual activity increased, and increased in exact proportion with bodily application, while freeing itself from every thing superfluous. It appeared that by dedicating to physical toil eight hours, that half of the day which I had formerly passed in the oppressive state of a struggle with ennui, eight hours remained to me, of which only five of intellectual activity, according to my terms, were necessary to me. For it appeared, that if I, a very voluminous writer, who had done nothing for nearly forty years except write, and who had written three hundred printed sheets;--if I had worked during all those forty years at ordinary labor with the working-people, then, not reckoning winter evenings and leisure days, if I had read and studied for five hours every day, and had written a couple of pages only on holidays (and I have been in the habit of writing at the rate of one printed sheet a day), then I should have written those three hundred sheets in fourteen years. The fact seemed startling: yet it is the most simple arithmetical calculation, which can be made by a seven-year- old boy, but which I had not been able to make up to this time. There are twenty-four hours in the day; if we take away eight hours, sixteen remain. If any man engaged in intellectual occupations devote five hours every day to his occupation, he will accomplish a fearful amount. And what is to be done with the remaining eleven hours?

It proved that physical labor not only does not exclude the possibility of mental activity, but that it improves its quality, and encourages it.

In answer to the question, whether this physical toil does not deprive me of many innocent pleasures peculiar to man, such as the enjoyment of the arts, the acquisition of learning, intercourse with people, and the delights of life in general, it turned out exactly the reverse: the more intense the labor, the more nearly it approached what is considered the coarsest agricultural toil, the more enjoyment and knowledge did I gain, and the more did I come into close and loving communion with men, and the more happiness did I derive from life.

In answer to the question (which I have so often heard from persons not thoroughly sincere), as to what result could flow from so insignificant a drop in the sea of sympathy as my individual physical labor in the sea of labor ingulfing me, I received also the most satisfactory and unexpected of answers. It appeared that all I had to do was to make physical labor the habitual condition of my life, and the majority of my false, but precious, habits and my demands, when physically idle, fell away from me at once of their own accord, without the slightest exertion on my part. Not to mention the habit of turning day into night and vice versa, my habits connected with my bed, with my clothing, with conventional cleanliness,--which are downright impossible and oppressive with physical labor,--and my demands as to the quality of my food, were entirely changed. In place of the dainty, rich, refined, complicated, highly-spiced food, to which I had formerly inclined, the most simple viands became needful and most pleasing of all to me,--cabbage-soup, porridge, black bread, and tea v prikusku. {3} So that, not to mention the influence upon me of the example of the simple working-people, who are content with little, with whom I came in contact in the course of my bodily toil, my very requirements underwent a change in consequence of my toilsome life; so that my drop of physical labor in the sea of universal labor became larger and larger, in proportion as I accustomed myself to, and appropriated, the habits of the laboring classes; in proportion, also, to the success of my labor, my demands for labor from others grew less and less, and my life naturally, without exertion or privations, approached that simple existence of which I could not even dream without fulfilling the law of labor.

It proved that my dearest demands from life, namely, my demands for vanity, and diversion from ennui, arose directly from my idle life. There was no place for vanity, in connection with physical labor; and no diversions were needed, since my time was pleasantly occupied, and, after my fatigue, simple rest at tea over a book, or in conversation with my fellows, was incomparably more agreeable than theatres, cards, conceits, or a large company,--all which things are needed in physical idleness, and which cost a great deal.

In answer to the question, Would not this unaccustomed toil ruin that health which is indispensable in order to render service to the people possible? it appeared, in spite of the positive assertions of noted physicians, that physical exertion, especially at my age, might have the most injurious consequences (but that Swedish gymnastics, the massage treatment, and so on, and other expedients intended to take the place of the natural conditions of man's life, were better), that the more intense the toil, the stronger, more alert, more cheerful, and more kindly did I feel. Thus it undoubtedly appeared, that, just as all those cunning devices of the human mind, newspapers, theatres, concerts, visits, balls, cards, journals, romances, are nothing else than expedients for maintaining the spiritual life of man outside his natural conditions of labor for others,--just so all the hygienic and medical devices of the human mind for the preparation of food, drink, lodging, ventilation, heating, clothing, medicine, water, massage, gymnastics, electric, and other means of healing,--all these clever devices are merely an expedient to sustain the bodily life of man removed from its natural conditions of labor. It turned out that all these devices of the human mind for the agreeable arrangement of the physical existence of idle persons are precisely analogous to those artful contrivances which people might invent for the production in vessels hermetically sealed, by means of mechanical arrangements, of evaporation, and plants, of the air best fitted for breathing, when all that is needed is to open the window. All the inventions of medicine and hygiene for persons of our sphere are much the same as though a mechanic should hit upon the idea of heating a steam-boiler which was not working, and should shut all the valves so that the boiler should not burst. Only one thing is needed, instead of all these extremely complicated devices for pleasure, for comfort, and for medical and hygienic preparations, intended to save people from their spiritual and bodily ailments, which swallow up so much labor,--to fulfil the law of life; to do that which is proper not only to man, but to the animal; to fire off the charge of energy taken win in the shape of food, by muscular exertion; to speak in plain language, to earn one's bread. Those who do not work should not eat, or they should earn as much as they have eaten.

And when I clearly comprehended all this, it struck me as ridiculous. Through a whole series of doubts and searchings, I had arrived, by a long course of thought, at this remarkable truth: if a man has eyes, it is that he may see with them; if he has ears, that he may hear; and feet, that he may walk; and hands and back, that he may labor; and that if a man will not employ those members for that purpose for which they are intended, it will be the worse for him.

I came to this conclusion, that, with us privileged people, the same thing has happened which happened with the horses of a friend of mine. His steward, who was not a lover of horses, nor well versed in them, on receiving his master's orders to place the best horses in the stable, selected them from the stud, placed them in stalls, and fed and watered them; but fearing for the valuable steeds, he could not bring himself to trust them to any one, and he neither rode nor drove them, nor did he even take them out. The horses stood there until they were good for nothing. The same thing has happened with us, but with this difference: that it was impossible to deceive the horses in any way, and they were kept in bonds to prevent their getting out; but we are kept in an unnatural position that is equally injurious to us, by deceits which have entangled us, and which hold us like chains.

We have arranged for ourselves a life that is repugnant both to the moral and the physical nature of man, and all the powers of our intelligence we concentrate upon assuring man that this is the most natural life possible. Every thing which we call culture,--our sciences, art, and the perfection of the pleasant thing's of life,-- all these are attempts to deceive the moral requirements of man; every thing that is called hygiene and medicine, is an attempt to deceive the natural physical demands of human nature. But these deceits have their bounds, and we advance to them. "If such be the real human life, then it is better not to live at all," says the reigning and extremely fashionable philosophy of Schopenhauer and Hartmann. If such is life, 'tis better for the coming generation not to live," say corrupt medical science and its newly devised means to that end.

In the Bible, it is laid down as the law of man: "In the sweat of thy face shalt thou eat bread, and in sorrow thou shalt bring forth children;" but "nous avons change tout ca," as Moliere's character says, when expressing himself with regard to medicine, and asserting that the liver was on the left side. We have changed all that. Men need not work in order to eat, and women need not bear children.

A ragged peasant roams the Krapivensky district. During the war he was an agent for the purchase of grain, under an official of the commissary department. On being brought in contact with the official, and seeing his luxurious life, the peasant lost his mind, and thought that he might get along without work, like gentlemen, and receive proper support from the Emperor. This peasant now calls himself "the Most Serene Warrior, Prince Blokhin, purveyor of war supplies of all descriptions." He says of himself that he has "passed through all the ranks," and that when he shall have served out his term in the army, he is to receive from the Emperor an unlimited bank account, clothes, uniforms, horses, equipages, tea, pease and servants, and all sorts of luxuries. This man is ridiculous in the eyes of many, but to me the significance of his madness is terrible. To the question, whether he does not wish to work, he always replies proudly: "I am much obliged. The peasants will attend to all that." When you tell him that the peasants do not wish to work, either, he answers: "It is not difficult for the peasant."

He generally talks in a high-flown style, and is fond of verbal substantives. "Now there is an invention of machinery for the alleviation of the peasants," he says; "there is no difficulty for them in that." When he is asked what he lives for, he replies, "To pass the time." I always look on this man as on a mirror. I behold in him myself and all my class. To pass through all the ranks (tchini) in order to live for the purpose of passing the time, and to receive an unlimited bank account, while the peasants, for whom this is not difficult, because of the invention of machinery, do the whole business,-- this is the complete formula of the idiotic creed of the people of our sphere in society.

When we inquire precisely what we are to do, surely, we ask nothing, but merely assert--only not in such good faith as the Most Serene Prince Blokhin, who has been promoted through all ranks, and lost his mind--that we do not wish to do any thing.

He who will reflect for a moment cannot ask thus, because, on the one hand, every thing that he uses has been made, and is made, by the hands of men; and, on the other side, as soon as a healthy man has awakened and eaten, the necessity of working with feet and hands and brain makes itself felt. In order to find work and to work, he need only not hold back: only a person who thinks work disgraceful- -like the lady who requests her guest not to take the trouble to open the door, but to wait until she can call a man for this purpose--can put to himself the question, what he is to do.

The point does not lie in inventing work,--you can never get through all the work that is to be done for yourself and for others,--but the point lies in weaning one's self from that criminal view of life in accordance with which I eat and sleep for my own pleasure; and in appropriating to myself that just and simple view with which the laboring man grows up and lives,--that man is, first of all, a machine, which loads itself with food in order to sustain itself, and that it is therefore disgraceful, wrong, and impossible to eat and not to work; that to eat and not to work is the most impious, unnatural, and, therefore, dangerous position, in the nature of the sin of Sodom. Only let this acknowledgement be made, and there will be work; and work will always be joyous and satisfying to both spiritual and bodily requirements.

The matter presented itself to me thus: The day is divided for every man, by food itself, into four parts, or four stints, as the peasants call it: (1) before breakfast; (2) from breakfast until dinner; (3) from dinner until four o'clock; (4) from four o'clock until evening.

A man's employment, whatever it may be that he feels a need for in his own person, is also divided into four categories: (1) the muscular employment of power, labor of the hands, feet, shoulders, back,--hard labor, from which you sweat; (2) the employment of the fingers and wrists, the employment of artisan skill; (3) the employment of the mind and imagination; (4) the employment of intercourse with others.

The benefits which man enjoys are also divided into four categories. Every man enjoys, in the first place, the product of hard labor,-- grain, cattle, buildings, wells, ponds, and so forth; in the second place, the results of artisan toil,--clothes, boots, utensils, and so forth; in the third place, the products of mental activity,-- science, art; and, in the forth place, established intercourse between people.

And it struck me, that the best thing of all would be to arrange the occupations of the day in such a manner as to exercise all four of man's capacities, and myself produce all these four sorts of benefits which men make use of, so that one portion of the day, the first, should be dedicated to hard labor; the second, to intellectual labor; the third, to artisan labor; and the forth, to intercourse with people. It struck me, that only then would that false division of labor, which exists in our society, be abrogated, and that just division of labor established, which does not destroy man's happiness.

I, for example, have busied myself all my life with intellectual labor. I said to myself, that I had so divided labor, that writing, that is to say, intellectual labor, is my special employment, and the other matters which were necessary to me I had left free (or relegated, rather) to others. But this, which would appear to have been the most advantageous arrangement for intellectual toil, was precisely the most disadvantageous to mental labor, not to mention its injustice.

All my life long, I have regulated my whole life, food, sleep, diversion, in view of these hours of special labor, and I have done nothing except this work. The result of this has been, in the first place, that I have contracted my sphere of observations and knowledge, and have frequently had no means for the study even of problems which often presented themselves in describing the life of the people (for the life of the common people is the every-day problem of intellectual activity). I was conscious of my ignorance, and was obliged to obtain instruction, to ask about things which are known by every man not engaged in special labor. In the second place, the result was, that I had been in the habit of sitting down to write when I had no inward impulse to write, and when no one demanded from me writing, as writing, that is to say, my thoughts, but when my name was merely wanted for journalistic speculation. I tried to squeeze out of myself what I could. Sometimes I could extract nothing; sometimes it was very wretched stuff, and I was dissatisfied and grieved. But now that I have learned the indispensability of physical labor, both hard and artisan labor, the result is entirely different. My time has been occupied, however modestly, at least usefully and cheerfully, and in a manner instructive to me. And therefore I have torn myself from that indubitably useful and cheerful occupation for my special duties only when I felt an inward impulse, and when I saw a demand made upon me directly for my literary work.

And these demands called into play only good nature, and therefore the usefulness and the joy of my special labor. Thus it turned out, that employment in those physical labors which are indispensable to me, as they are to every man, not only did not interfere with my special activity, but was an indispensable condition of the usefulness, worth, and cheerfulness of that activity.

The bird is so constructed, that it is indispensable that it should fly, walk, peek, combine; and when it does all this, it is satisfied and happy,--then it is a bird. Just so man, when he walks, turns, raises, drags, works with his fingers, with his eyes, with his ears, with his tongue, with his brain,--only then is he satisfied, only then is he a man.

A man who acknowledges his appointment to labor will naturally strive towards that rotation of labor which is peculiar to him, for the satisfaction of his inward requirements; and he can alter this labor in no other way than when he feels within himself an irresistible summons to some exclusive form of labor, and when the demands of other men for that labor are expressed.

The character of labor is such, that the satisfaction of all a man's requirements demands that same succession of the sorts of work which renders work not a burden but a joy. Only a false creed, [Greek text which cannot be reproduced], to the effect that labor is a curse, could have led men to rid themselves of certain kinds of work; i.e., to the appropriation of the work of others, demanding the forced occupation with special labor of other people, which they call division of labor.

We have only grown used to our false comprehension of the regulation of labor, because it seems to us that the shoemaker, the machinist, the writer, or the musician will be better off if he gets rid of the labor peculiar to man. Where there is no force exercised over the labor of others, or any false belief in the joy of idleness, not a single man will get rid of physical labor, necessary for the satisfaction of his requirements, for the sake of special work; because special work is not a privilege, but a sacrifice which man offers to inward pressure and to his brethren.

The shoemaker in the country, who abandons his wonted labor in the field, which is so grateful to him, and betakes himself to his trade, in order to repair or make boots for his neighbors, always deprives himself of the pleasant toil of the field, simply because he likes to make boots, because he knows that no one else can do it so well as he, and that people will be grateful to him for it; but the desire cannot occur to him, to deprive himself, for the whole period of his life, of the cheering rotation of labor.

It is the same with the starosta [village elder], the machinist, the writer, the learned man. To us, with our corrupt conception of things, it seems, that if a steward has been relegated to the position of a peasant by his master, or if a minister has been sent to the colonies, he has been chastised, he has been ill-treated. But in reality a benefit has been conferred on him; that is to say, his special, hard labor has been changed into a cheerful rotation of labor. In a naturally constituted society, this is quite otherwise. I know of one community where the people supported themselves. One of the members of this society was better educated than the rest; and they called upon him to read, so that he was obliged to prepare himself during the day, in order that he might read in the evening. This he did gladly, feeling that he was useful to others, and that he was performing a good deed. But he grew weary of exclusively intellectual work, and his health suffered from it. The members of the community took pity on him, and requested him to go to work in the fields.

For men who regard labor as the substance and the joy of life, the basis, the foundation of life will always be the struggle with nature,--labor both agricultural and mechanical, and intellectual, and the establishment of communion between men. Departure from one or from many of these varieties of labor, and the adoption of special labor, will then only occur when the man possessed of a special branch, and loving this work, and knowing that he can perform it better than others, sacrifices his own profit for the satisfaction of the direct demands made upon him. Only on condition of such a view of labor, and of the natural division of labor arising from it, is that curse which is laid upon our idea of labor abrogated, and does every sort of work becomes always a joy; because a man will either perform that labor which is undoubtedly useful and joyous, and not dull, or he will possess the consciousness of self- abnegation in the fulfilment of more difficult and restricted toil, which he exercises for the good of others.

But the division of labor is more profitable. More profitable for whom? It is more profitable in making the greatest possible quantity of calico, and boots in the shortest possible time. But who will make these boots and this calico? There are people who, for whole generations, make only the heads of pins. Then how can this be more profitable for men? If the point lies in manufacturing as much calico and as many pins as possible, then this is so. But the point concerns men and their welfare. And the welfare of men lies in life. And life is work. How, then, can the necessity for burdensome, oppressive toil be more profitable for people? For all men, that one thing is more profitable which I desire for myself,-- the utmost well-being, and the gratification of all those requirements, both bodily and spiritual, of the conscience and of the reason, which are imposed upon me. And in my own case I have found, that for my own welfare, and for the satisfaction of these needs of mine, all that I require is to cure myself of that folly in which I had been living, in company with the Krapivensky madman, and which consisted in presupposing that some people need not work, and that certain other people should direct all this, and that I should therefore do only that which is natural to man, i.e., labor for the satisfaction of their requirements; and, having discovered this, I convinced myself that labor for the satisfaction of one's own needs falls of itself into various kinds of labor, each one of which possesses its own charm, and which not only do not constitute a burden, but which serve as a respite to one another. I have made a rough division of this labor (not insisting on the justice of this arrangement), in accordance with my own needs in life, into four parts, corresponding to the four stints of labor of which the day is composed; and

I seek in this manner to satisfy my requirements.

These, then, are the answers which I have found for myself to the question, "What is to be done?"

First, Not to lie to myself, however far removed my path in life may be from the true path which my reason discloses to me.

Second, To renounce my consciousness of my own righteousness, my superiority especially over other people; and to acknowledge my guilt.

Third, To comply with that eternal and indubitable law of humanity,- -the labor of my whole being, feeling no shame at any sort of work; to contend with nature for the maintenance of my own life and the lives of others.

Footnote:

{1} An omission by the censor, which I am unable to supply. TRANS.

{2} We designate as organisms the elephant and the bacterian, only because we assume by analogy in those creatures the same conjunction of feeling and consciousness that we know to exist in ourselves. But in human societies and in humanity, this actual sign is absent; and therefore, however many other signs we may discover in humanity and in organism, without this substantial token the recognition of humanity as an organism is incorrect.

{3} v prikusku, when a lump of sugar is held in the teeth instead or being put into the tea.

.

CPSIA information can be obtained
at www.ICGtesting.com
Printed in the USA
BVHW040319230419
546159BV00027BA/2200/P